Harrison Allen

Report of an Autopsy on the Bodies of Chang and Eng Bunker

Commonly Known as the Siamese Twins

Harrison Allen

Report of an Autopsy on the Bodies of Chang and Eng Bunker
Commonly Known as the Siamese Twins

ISBN/EAN: 9783337248963

Printed in Europe, USA, Canada, Australia, Japan

Cover: Foto ©berggeist007 / pixelio.de

More available books at **www.hansebooks.com**

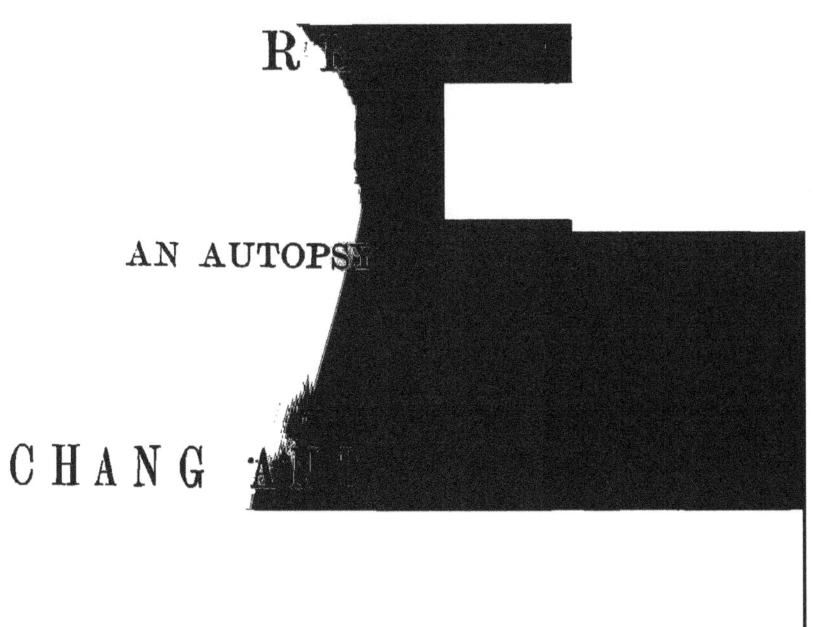

BY

HARRISON ALLEN, M.D.,

PROFESSOR OF COMPARATIVE ANATOMY AND ZOÖLOGY IN THE UNIVERSITY OF PENNSYLVANIA, SURGEON TO THE PHILADELPHIA HOSPITAL, ETC.

PHILADELPHIA:

COLLINS, PRINTER, 705 JAYNE STREET.

1875.

AUTOPSY OF THE SIAMESE TWINS.

Fig. 1.

Fig. 1. The twins in the acquired position (E. R., C. L.). From a photograph taken in St. Petersburg, 1870.

REPORT

OF AN

AUTOPSY ON THE BODIES OF CHANG AND ENG BUNKER, COMMONLY KNOWN AS THE SIAMESE TWINS

NOTE.

The word "Report" used in the title of my paper is to be read as referring to the post-mortem appearances only, and not to the Report of the Commission as appointed by the College. The ante-mortem history prefixed to my paper was written in conjunction with Prof. Pancoast.

H. ALLEN.

They were married in April, 1843, and raised large families; Chang having had ten, and Eng twelve children. Chang had three boys and seven girls; Eng had seven boys and five girls. These were in all respects average children, excepting two, a boy and girl of Chang's, who were deaf-mutes.

The twins resided in a rolling country, about four miles from Mount Airy, Surrey Co., N. C. They were prosperous farmers, each owning his own farm. The

[1] For this statement see an article in Lippincott's Magazine, March, 1874.

Fig. 1.

Fig. 1. The twins in the acquired position (E. R., C. L.). From a photograph taken in St. Petersburg, 1870.

REPORT

OF AN

AUTOPSY ON THE BODIES OF CHANG AND ENG BUNKER, COMMONLY KNOWN AS THE SIAMESE TWINS.

By

HARRISON ALLEN, M.D.,

PROFESSOR OF COMPARATIVE ANATOMY AND ZOÖLOGY IN THE UNIVERSITY OF PENNSYLVANIA, SURGEON TO THE PHILADELPHIA HOSPITAL, ETC.

[Read April 1, 1874.]

ANTE-MORTEM HISTORY.

Chang and Eng Bunker were born near Bangkok, Siam, in 1811, their father being a Chinaman, their mother a native of Siam, bred by a Chinese father.[1]

The twins were united by a band extending from the junction of the abdominal and thoracic cavities, anteriorly, constituting the variety in teratology known as *Omphalopagus xiphodidymus.*

They were married in April, 1843, and raised large families; Chang having had ten, and Eng twelve children. Chang had three boys and seven girls; Eng had seven boys and five girls. These were in all respects average children, excepting two, a boy and girl of Chang's, who were deaf-mutes.

The twins resided in a rolling country, about four miles from Mount Airy, Surrey Co., N. C. They were prosperous farmers, each owning his own farm. The

[1] For this statement see an article in Lippincott's Magazine, March, 1874.

dwellings of the two families were a mile and a half apart. The twins resided three days in each of the homes alternately. They were expert in the handling of tools, in plowing, shingling, shooting, etc. They lived much in the open air, and frequently drove in a carriage to the neighboring village.

The events leading to their death were as follows: About six years ago Chang, who had always been the more excitable, became addicted to immoderate drinking. Three years ago, while on a voyage from Liverpool to New York, he was stricken with hemiplegia of the right side. He in great measure recovered from this attack, but could never ascend and descend stairs with facility. For this reason the twins occupied rooms on the ground floors of their homes.

On Monday evening, Jan. 12th, 1874, Chang was seized, while at his own house, with an attack of bronchitis. He had a cough; scanty, frothy sputa; but no pain. On Wednesday the symptoms had somewhat subsided; the skin was acting freely. Loud bronchial râles were present over the left side of the chest. On Thursday evening the twins insisted upon leaving Chang's house for Eng's. The weather was very cold, and the journey was undertaken in an open carriage. On their arrival, however, Chang continued as well as before, until Friday evening, when he complained of thoracic oppression and inability to lie down with comfort. After having retired that evening, the twins were heard to get up, and go out on the porch, by the side of the house, where they drank of water, and returned to their room. They built a large wood fire, and sat down; Eng soon complaining of being sleepy, Chang declaring that he could not breathe if he should lie down. Finally they again retired. They both fell asleep. Near day-break (Jan. 17th) Eng called to one of his sons, who slept in a room above, to come down and waken Chang. The boy soon made his appearance, and going to the side of Chang, cried out, "Uncle Chang is dead!" Eng at once said "Then I am going!"—It is probable that Chang was sleeping when he died.

Eng made no further mention of Chang other than to request that the body be moved closer to him. Soon afterward Eng desired to have his limbs moved. This desire continued for half an hour. He then asked for a urinal, but did not void over a few drops of urine. He several times repeated the endeavor to micturate, but without success. He then complained of a choking sensation, and asked to be raised in bed. He had continued rational. His last words were "May God have mercy on my soul!" He gradually became fainter, fell into a syncopal state, and died quietly a little over two hours from the announcement to him of the death of his brother.

AUTOPSY.

The Autopsy was begun in the house of Eng, Sunday, February 1st, 1874, and finished in the Mütter Museum of the College of Physicians, at Philadelphia.

Age of subjects, 63 years. Examination made about fifteen days and eight hours after death. The weather had been cold. No preservative had been employed prior to the date of the autopsy.

I. Post-mortem Appearances.

The following is their description in Chang.

Body moderately emaciated. *Rigor mortis* none. The fingers of the right hand were semi-flexed, a condition due doubtless to the long-standing paralysis of the right upper extremity. Passive congestion was marked over entire dorsal aspect of the neck, trunk, and upper extremities. It was less marked over the corresponding surfaces of the forearm and legs. The feet and hands were almost entirely free. The superficial veins in the last-named localities, especially in the left foot, were distended. The passive congestion extended over the right thoracic region as far as the median line, and on the front of both thighs, especially the right. Upon the head the congestion was seen behind and beneath the ears, and was sparsely distributed over both malar prominences. The lips were discolored. The integument of the genitals was infiltrated, the scrotum particularly being much swollen. There was extensive greenish discoloration on the anterior abdominal wall. The left external abdominal ring was enlarged. Both testicles were within the scrotum. The hair of the head was gray. That on the right side of the pubis was black, that on the left was of an iron-gray color.

The following is their description in Eng.

Body moderately well nourished, spare. *Rigor mortis* slight. Passive congestion less marked than in Chang. It was most conspicuous on the buttocks and infraspinous spaces. There was none in front of the body. The testicle of left side absent from scrotum. There was moderate greenish discoloration of anterior wall of abdomen. The hair on the pubis was black on the right side, pure gray on left side of the median line.

Measurements.—Chang was 5 ft. 2½ in. in height; Eng, 5 ft. 3½ in. When the bodies were laid upon a table Chang's left side and Eng's right side were drawn somewhat toward one another. This was most marked in Chang, and gave a greater inclination of his trunk toward his brother's. (See Figs. 1 and 19.)

II. External Appearances of the Band.

When the bodies were suspended and placed face to face, it was evident that the congenital position had been secured. All observations were made, as far as was possible, with the bodies in this position. With it the details of structure, it was thought, could be easily understood; without it the subject would be difficult and confused.

The "band" was a massive commissure placed between the bodies at the junction of the abdominal and thoracic regions of each. It was broader above than below, and had a circumference of nine inches. It presented four surfaces for examination, an upper, lower, and two lateral surfaces.

The *upper* surface was somewhat flattened in both Chang and Eng. The ensiform cartilage of each body could be felt deflected from the sternum and prolonged into the band. The base of the cartilage in Eng presented a rounded circular eminence, measuring one inch in diameter. There was no corresponding eminence in Chang. The upper aspect of each process could be well defined beneath the skin, the subcutaneous connective tissue being more abundant in Chang than in Eng. The upper surface measured $2\frac{1}{2}$ in. in width at its base towards Eng, and $2\frac{1}{3}$ in. at its base towards Chang. It was 2 in. wide at its middle.

Fig. 2.

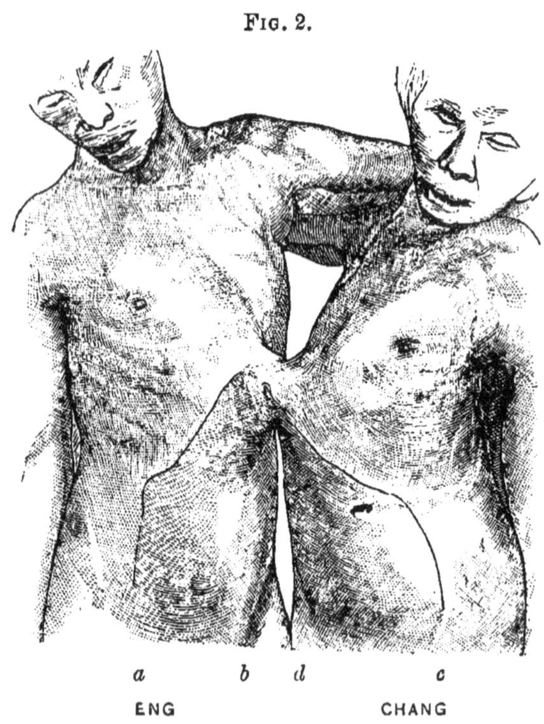

Fig. 2. The twins in the acquired position (E. R., C. L.), showing band and the primary incisions, *a–b*, *c–d*. From a photograph taken after death at Philadelphia.

The *lower* surface was much narrower than the upper. It was marked in the centre, but nearer the anterior than the posterior border, by a linear scar one inch in length, which it was thought answered to the position of the single umbilicus. The skin was adherent at this point, but elsewhere was easily raised in folds. Behind the scar, *i. e.*, toward the posterior part of the band, the skin was somewhat corrugated. This portion answered, in position, to Chang's umbilical pouch.

The *lateral* surfaces. The terms *upper* surface and *lower* surface have fixed values, no matter how they may be approached by the observer. This is not the case, however, with the lateral surfaces, as will appear from the following considerations. Viewing the band as a separate form—as it was spoken of during the life of the twins—we will see that the terms front ("anterior") and back ("posterior"), as given to the lateral surfaces, were derived from studying the acquired position. Thus we were bound not to cut the "front" of the band, but allowed to make an incision on the "back." Now this position of selection was destroyed, and its terms deprived of what meaning they may have had, by the reproduction of the congenital relations of the bodies.

There is no doubt that in infancy and early childhood there was no acquired position, and, therefore, neither "front" nor "back" to the band. And later, when, as we have reason to believe, the position of selection was gradually adopted, the terms "front" and "back" were reversible—the "front" meaning that which corresponded to the surface of least thoracic approximation. Thus when the adult condition was fixed, and the "front" answered to the widely separated right side of Eng's chest and left side of Chang's chest, the "back" was in relation with the closely approximated left side of Eng's chest and right side of Chang's.

To avoid awkward repetition of phrases expressing the facts of the last sentence, the following characters will be employed in describing the "lateral" surfaces of the band.

E. R., C. L. (Eng's right, Chang's left) will designate the "anterior" surface of the acquired position. C. R., E. L. (Chang's right, Eng's left) will designate the "posterior" surface. Since the right side of Chang's half of the band merged into the left on Eng's half, while the right side of Eng's half, after it passed the middle line, became the left half of Chang's, we propose using the characters E. L., E. R., and C. R., C. L., which will be understood as signifying left side Eng, right side Eng, etc.

Using the above signs we found that the surface E. R., C. L. was inclined decidedly downward and backward when seen in the congenital position, and was 3 in. high and $2\frac{1}{2}$ in. wide. At E. R., the border answering to the ensiform cartilage was marked by a large rounded tubercle; it was much more pronounced than on the corresponding border of C. L. When seen in the acquired position, E. R., C. L. became "anterior," when, at its upper margin, C. L. was longer than E. R. by one-half inch. The tubercle on E. R., already noticed, was much more prominent than C. L. The contour of the inferior margin was also different, being more uneven. C. L. was not only longer, but was more obliquely placed downward and outward to the centre of the band than E. R. (See cast in the Mütter Museum.)

III. Coverings of the Band.

In front (E. R., C. L.) the superficies could not be well examined owing to the restrictions imposed by the families. A view of it from within can be obtained in

Figs. 8, 9, *q. v.* Permission having been granted to make an incision "behind," at C. R., E. L., a modified letter-H incision was employed, thus—

Turning the skin flaps here indicated upwards and downwards, and the lateral triangles outwards, there was found beneath, a layer of superficial fatty connective tissue, with a well-defined layer of fat on either side, but with an intermediate portion which was free from fat, and of greater thickness.

The skin could with some little trouble be raised over the *dorsal or upper surface*, showing here entire absence of fat. A very delicate artery was found running across the middle, from Eng to Chang.

The lower portion of the surface C. R., E. L. was inseparably connected with the umbilicus. It was also united to the superficial fascia on C. L., about $1\frac{1}{2}$ inches from the umbilicus; this did not have any connection with the deeper parts. The process of fibrous tissue which had been felt through the skin was conspicuous on this surface of the band, and was covered by a delicate non-fatty layer of connective tissue. Towards the lower part of the surface were seen several diverging lines of fibrous tissue, which were lost within the integument about the umbilical scar, at the lower surface of the band. They were exceedingly thin, and at one point lay directly over the posterior and inferior wall of the umbilical pouch of Chang.

Turning down the superficial layer, the aponeurosis of the external oblique muscle was exposed (Fig. 3). A marked contrast was exhibited in the two sides of the band. In Chang the parts were normal so far as they were exposed—the characteristic apertures for the escape of small vessels being abundant and conspicuous. In turning down the superficial fascia in Eng (Fig. 3, A), it was found to be continuous at its

FIG. 3.

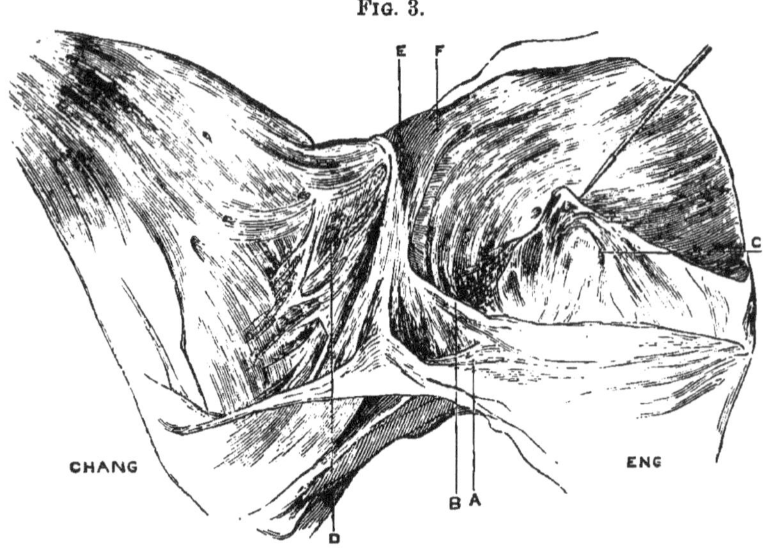

FIG. 3. The surface, C. R., E. L., exposed by removal of skin and superficial fascia to display the tendons of the external oblique muscles and adjacent parts.

A. The superficial fascia—lost over the position of Chang's umbilical pouch.

B, C. Supplemental layers of fibrous tissue of Eng not seen in Chang; B is a continuation toward Eng of aponeurotic fibres having a source from the linea alba of Chang; C is independent of the former, and is continuous with the deep pectoral fascia.

D. The interlacing of fibres on tendon of external oblique muscle of Chang.

E. The linea alba of Chang, beginning at C. R.

F. Its continuation to E. L., and insertion upon the ensiform cartilage.

lower portion with an aponeurotic layer (Fig. 3, B), which extended toward the median line, where it was continuous with the linea alba of Chang. In addition to this, a second layer (Fig. 3, C), analogous in position to a deep layer of the superficial fascia, which was entirely independent of Chang, extended over nearly the whole of Eng's division, and was particularly well defined over the fibres of origin of the external oblique muscle. This was continuous with the deep layer of the superficial fascia which passed over the entire side of Eng's thorax.

Toward the middle of the band this layer gradually lost its distinctive features, and was firmly incorporated with the tendon of the external oblique muscle. A number of fibres corresponding to it extended in inseparable intimacy with this tendon. These were gradually lost as they approached the linea alba of Chang, and the parts being in position these fibres were at their lower portion covered in by the aponeurotic extension of the linea alba already mentioned.

On Chang's side, as we have seen, the parts comparable to these accessory layers were absent. There was no line of demarcation between the tendon of the oblique and the aponeurotic attachment of the pectoralis. The tendon of the external oblique presented a different appearance from the normal one in a more extensive interlacing of fibres of the linea alba with the tendon.

The part termed above the linea alba of Chang (Fig. 3, E), has already been indicated through skin and superficial fascia. As can be seen, this band of fibres, having its origin from the middle line of the abdomen of Chang, was found to be a direct continuation of the linea alba. It was remarkable in not being inserted into the ensiform process of Chang, but into that of Eng, and in giving off the aponeurotic outshoot B, already noticed, as well as in having a diffused point of insertion into Eng's tissue as in the ensiform cartilage (F). In a word, the linea alba approaches the surface C. R., E. L. from C. R. below, and is inserted into E. L. above.

IV. Organs of Abdomen as observed in position through the Incisions.

Limited incisions being alone permitted, the large vessels of the abdomen were sought for in the process of embalmment, believing, as we did, that the procedures of securing them would enable us, by extending the cuts from below upward, to fairly open the abdomen and examine thereby the interior of the band.

In each body, therefore, an incision six inches long (Fig. 2, *a b*, *c d*) was extended from the centre of the right iliac region to the centre of the right hypochondriac region. This was subsequently joined by an oblique incision passing from the upper end of the first mentioned to the lateral border of the ensiform cartilage at its base. This incision measured $7\frac{1}{2}$ in. The lower end of the vertical incision was met by a horizontal one passing to the centre of the hypogastric region, and measuring $3\frac{1}{4}$ in.

Through these incisions were studied (1) the *umbilical ligaments* and (2) *the abdominal viscera.*

1. *The umbilical ligaments.*[1]—By turning forward the anterior flap in Eng as far as possible, the peritoneal lining was exhibited, and there was brought into view a structure beginning at the summit of the bladder, and which, ascending the abdominal wall and passing obliquely to the right side, could be traced

[1] The folds of peritoneum containing remains of the hypogastric arteries will be called throughout by the name of *the umbilical ligaments.*

clearly to the scar-like tissue marking the remains of the umbilical structures situated upon the anterior abdominal wall within about $1\frac{1}{2}$ in. of the band. This structure was the umbilical ligament (Fig. 4, A). It was loaded with fat, and, as it terminated at the scar, distinct lobules of fat (several of which were pedunculated) were abundantly deposited.

The bladder was distended and raised 5 in. above the pubis.

FIG. 4.

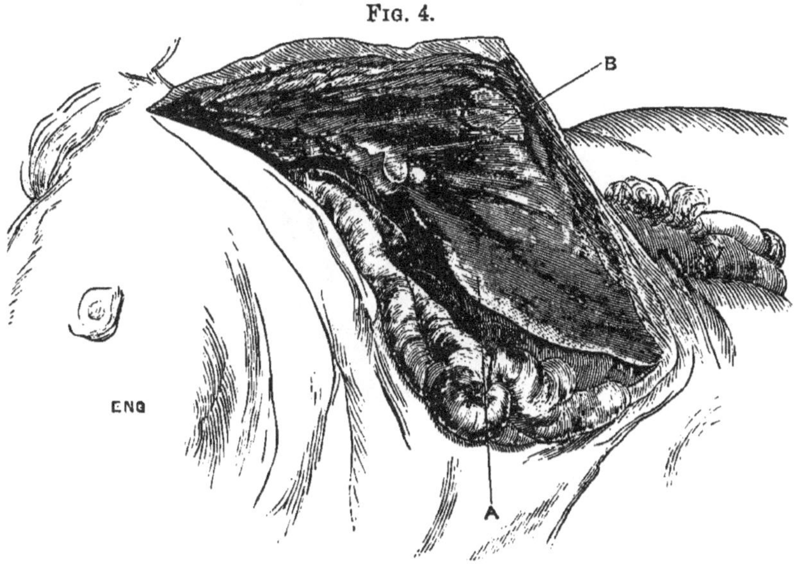

FIG. 4. The umbilical ligament in Eng.
A. The umbilical ligament.
B. The lobule of fat at position of the normal umbilicus.

In Chang (Fig. 5), the same appearances were seen as those above given, save that no fat was deposited in the umbilical ligament. On the contrary, it resembled the omentum of an emaciated subject. When stretched, the fold was fully an inch wide, quite transparent, and marked by two longitudinal bands, which recalled the shapes of the obliterated vessels. But two rather small sessile fatty appendages were seen at the scar.

The bladder was empty, contracted, and lay within the true pelvis.

FIG. 5.

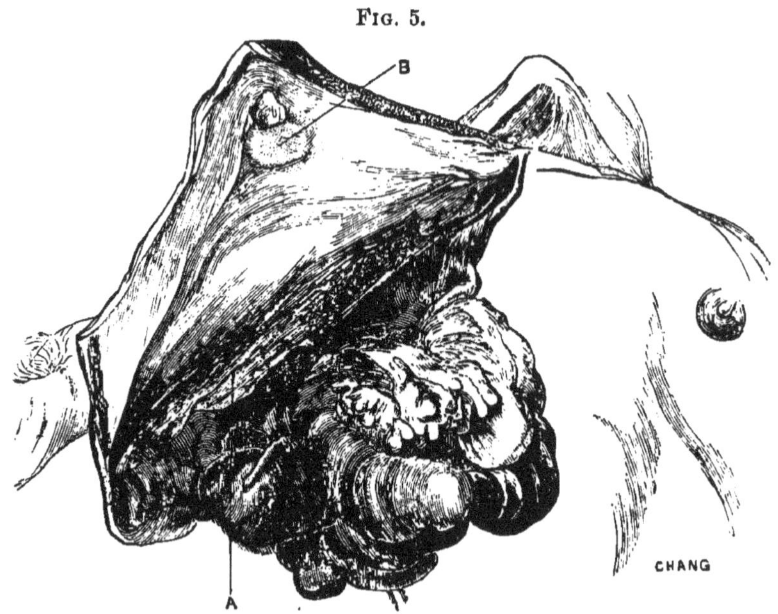

FIG. 5. The umbilical ligament in Chang.
A. The umbilical ligament.
B. The lobule of fat at position of the normal umbilicus.

In both Chang and Eng an isolated mass of sub-peritoneal fat, presenting a sub-circular form, and measuring 1 in. in diameter, was found in the position of the normal umbilicus (Figs. 4 and 5).

2. *The viscera.*—In Eng the omentum was gathered up toward the transverse colon. It was abundantly furnished with fat.[1] The transverse colon extended

[1] The presence of a great amount of adipose tissue throughout, in Eng, was very noticeable as contrasted with the emaciated appearance of the tissues in Chang.

across the abdomen, beginning on the right side on a level with the eleventh rib. It was contracted and contained a little flatus. The rest of the exposed region was occupied by coils of small intestine, yielding a mesentery very rich in fat. The stomach was not visible. By removing the small intestine, and bringing down the transverse colon and large intestine, the pyloric extremity of the stomach was seen. The fundus of stomach, spleen, and left kidney were not seen. (Fig. 6.)

FIG. 6.

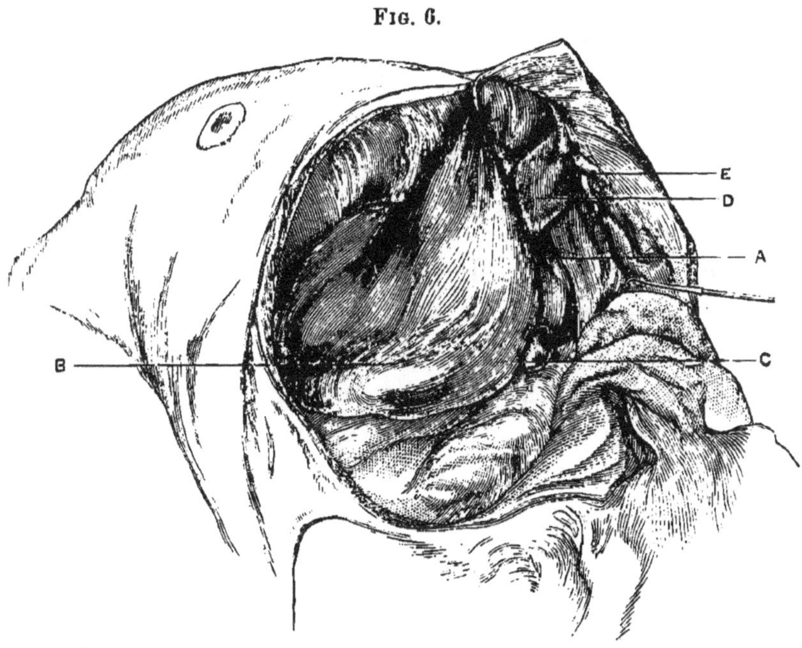

FIG. 6. The abdominal organs in Eng—the small intestines removed.

A. Left lobe of liver.

B. Right lobe of liver.

C. Gall-bladder.

D. Suspensory ligament.

E. Lobules of fat in the position of the termination of the umbilical ligament.

The liver.—The right lobe was alone visible. This extended entirely across the right hypochondriac and epigastric regions. Its external free border was not in contact with the ribs. Between it and the external abdominal wall there was an interval of nearly an inch at its greatest part, which was crossed by the external lateral ligament. The inferior border of the lobe rested upon and nearly concealed the pylorus of the stomach as well as the upper half of right kidney. Corresponding in position to the upper portion of the right kidney was a well-defined layer of peritoneum, presenting a sharply defined internal border. Upon dissecting away the peritoneum from this border it was found to answer to the inferior vena cava. The lesser omentum occupied its usual position. The fundus of the gall-bladder was two-thirds of an inch beyond the anterior border of the lobe, immediately to the outer side of the caudal lobe. The position of the longitudinal fissure was well off to the left side of the abdomen, presenting, between the right and left lobes, a conspicuous cleft which was partially occupied by the base of the caudal lobe. The round ligament, with its associated suspensory ligament, had doubtless passed nearly vertically, before the relations had been disturbed by the incision in the abdominal wall, upwards and forwards to the anterior abdominal wall at a point lying one inch to the outer side of the centre of the umbilicus.

In the subject, as it lay on the table with the flap *a*, *b* (Fig. 2), turned to the left, the suspensory ligament had the appearance of being much more obliquely inclined to the left, and could be made nearly horizontal by a little traction. Lying beneath this ligament, but belonging to the anterior abdominal wall, was a large mass of subperitoneal fat about the size of a pigeon's egg. Extending to the extreme left, and continuous with the anterior border of the left lobe of the liver, was a delicate prolongation of liver substance which was lost within the connecting band.

FIG. 7.

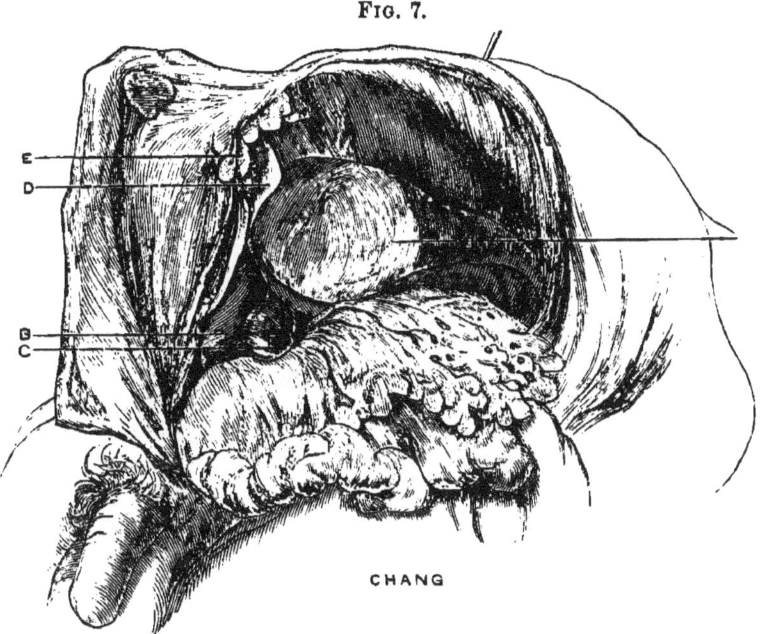

FIG. 7. The abdominal organs of Chang in position—the small intestines removed.

A. Left lobe of liver.
B. Right lobe of liver.
C. Gall-bladder.
D. Suspensory ligament.
E. Lobules of fat in the position of the termination of the umbilical ligament.

The upper surface of this prolongation was supported by a fold of peritoneum, extending directly upward, apparently attached to the base of the ensiform cartilage. Visible upon the anterior aspect of this fold was a tortuous artery, afterwards found to be the left internal mammary. This fold may be called *the accessory suspensory ligament;* nothing similar to it was seen in Chang. The left lobe of the liver, save a portion of its anterior edge, was not visible.

In Chang (Fig. 7), by exposing the parts as in Eng, throwing the abdominal flap, *c*, *d* (Fig. 2), to the right, there was at once brought into view the transverse colon, the greater omentum, and greater curvature of the stomach. The latter organ was large, empty, and without any undue traction could be so displayed as to yield its fundus and greater curvature in position. The fundus was not visible. Lying conspicuously within the left hypochondriac region was the spleen. Its inferior free border, with its peritoneal attachment, was distinctly seen; its upper portions, however, were invisible. The left lobe of the liver held a position answering to that of the right lobe in Eng—the external lateral ligament being stretched across the left hypochondriac region, pursuing a similar course to the external lateral ligament of Eng (*q. v.*). The left lobe at its outer portion rested upon the spleen, its inner portion upon the stomach. The outer portion of the left lobe presented a thin compressed border, the inner portion was divided by a deep sulcus into two lobes.

The right lobe lay deep within the right hypochondriac region, the portion about the longitudinal fissure anteriorly, alone appearing in the dissection. The suspensory ligament held a position similar to that in Eng. The gall-bladder held its normal position to the right lobe, and was moderately distended with bile.

Both Chang and Eng had the organs occupying the hypochondriac and epigastric regions retaining, on the whole, such relations as are usually observed.

This statement appears pertinent, at this stage of the autopsy, in order to explain—

V. Interior of Band.

We here describe (1) *the hepatic pouches;* (2) *the umbilical pouches;* (3) *the vascular structures of the band;* (4) *the diaphragms;* (5) *the ensiform cartilages.*

1. *The hepatic pouches.*—The photograph (Fig. 2) indicates the position of the right lobe of Eng's liver in the right hypochondriac region. The right lobe of Chang is of course not seen in the figure, since it lies on the side of the body which is not in the field of vision. It must follow from the rights and lefts of the two individuals being opposites that, in drawing a line between the livers (which, as already seen, occupy normal positions as to right and left) across the band, such a line will be diagonal to the axes of the ensiform cartilages; Chang's half of the band having the line enter the band from his "right," Eng from his "left." It will also follow that any pouches of peritoneum which might accompany this line will pursue a similar direction—be on the same plane—be right or left with respect to the axis of that plane. Now it was actually demonstrated that such a line did extend between the livers, and was accompanied by such peritoneal pouches. These pouches were termed the hepatic pouches, and may be described as follows:—

Chang.—The subject lying on the table with rights and lefts determined as in the acquired position, the finger could be inserted behind the suspensory ligament (Fig. 7) in a pouch lying directly beneath the ensiform cartilages, into which passed an extension of liver-like tissue.

Eng.—This fact could not be well demonstrated in Eng in this position, but is well seen in Fig. 8.

It follows that the two hepatic pouches are on nearly the same plane, and that each approaches the central point of the band diagonally from the right side of the subject with whose abdominal cavity it is continuous.

2. *The umbilical pouches.*—Beneath the hepatic pouches, and between them and the inferior border of the band, were two pouches which, from their association with the round ligament, have been termed *the umbilical pouches.*

When the finger was passed toward the band from the abdomen of Chang, and following the peritoneum of the anterior wall of the abdomen, it passed into a pouch of the band directly over the skin covering, across the band, over the umbilicus, and was received within the folds of the suspensory ligament of the liver of Eng. This pouch was so superficial that while the finger was in the pouch any motion of the finger was readily followed by the observer.

In the same way as above, if the finger was introduced *behind* the suspensory ligament of Eng, it slipped into a pouch which passed across the median line of the band, and was received within the folds of the suspensory ligament of the liver of Chang.

There were then two pouches communicating with the two abdominal cavities, arranged one above another in the band, Chang's being the lower of the two. No remains of an umbilical vein were detected, nor was there any communication between the pouches and the umbilicus. It has already been noticed that the round ligament of each liver passed from the longitudinal fissure to a scar on the anterior wall of the abdomen near the band. It was not, therefore, within the round ligaments, but the folds of the suspensory ligaments, that the pouches were found.

Eng's pouch measured 2½ in. From edge of Chang's suspensory ligament to end of hepatic pouch measured 3 in.

Extending across the band, about midway between the properties of the two individuals, was a septum. It was attached above and below to the respective boundaries of the band, and along its entire length was incorporated with its two peritoneal cavities, so that when in the course of the dissection of the "posterior" surface of the band the peritoneal covering of the band was displayed, several large lobules of fat were seen lying to Eng's side of the septum.[1]

It will be seen that Fig. 8 represents the band opened to display the pouches with the septum. The lower end of the septum is fixed near the scar of the umbilicus, and holds an immobile position over the umbilical pouches. At this point it is free from fat. But as it extends over the hepatic pouches it is more pliant. This portion of the septum has been carried a little to Chang's side of the band to display the entire length of the hepatic pouch of Eng.

[1] Before the septum was known to exist, the band was opened from behind in the presence of the Fellows of the College (Feb. 18th, 1874). The exact relations of the septum could not at that time be determined. Figs. 8, 9, and 10 are taken from studies of the parts made the day after the meeting.

Figs. 9 and 10 are designed to exhibit the appearances presented in securing views of the septum from

FIG. 8.

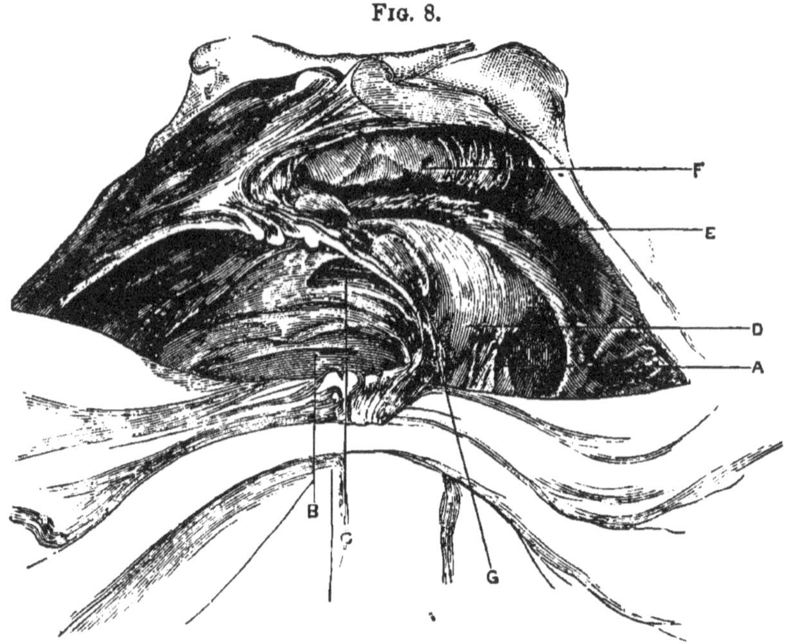

FIG. 8. The surface, C. R., E. L., showing the interior of band by free division of the aponeuroses seen in Fig. 7, and their underlying peritoneal attachments.

A. The orifice of umbilical pouch of Eng.
B. The orifice of umbilical pouch of Chang, showing connection with suspensory ligament of Eng.
C. The fenestrated umbilical pouch of Eng passing between the folds of the suspensory ligament of Chang.
D. Suspensory ligament of liver of Eng.
E. Hepatic tract.
F. Hepatic pouch of Eng.
G. The septum.

its sides. Fig. 9 is the side toward Chang, and Fig. 10 is the side toward Eng.

The pouches and septum were now removed and the position of the hepatic tract determined. It rested upon the incurved borders of the ensiform cartilages (see Fig. 15), and as the subject lay on the table with the "posterior" surfaces of the band exposed (Fig. 11) the hepatic tract was slightly arched. It measured three inches in length, was compressed, and measured six lines wide and three lines thick. The tract arose from the livers at the same point—namely, directly

FIG. 9.

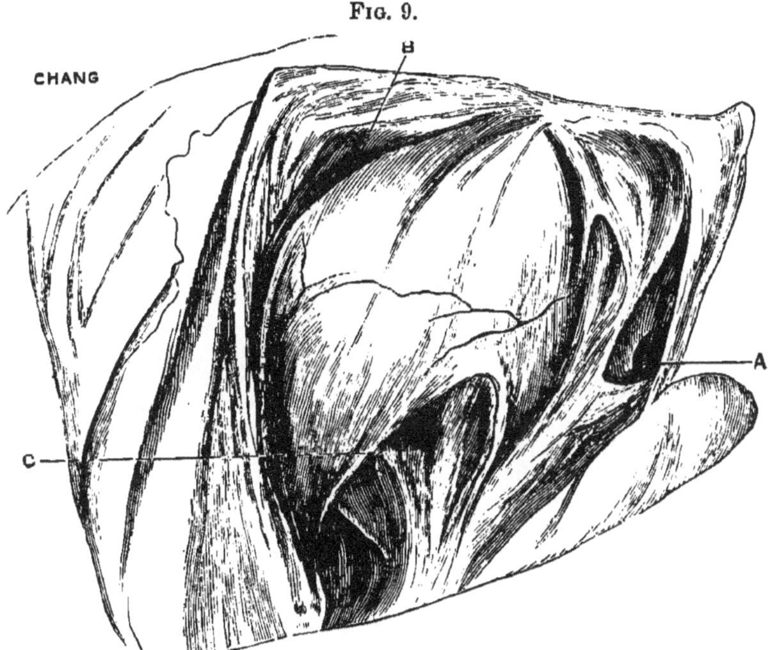

FIG. 9. The septum viewed from Chang's side.
A. The orifice of umbilical pouch of Chang.
B. The orifice of the hepatic pouch of Chang.
C. Suspensory ligament of Chang containing umbilical pouch of Eng.

above the longitudinal fissure, having more fulness on the side of the right than of the left lobe. The round ligament, as it passed out of the longitudinal fissure of each liver, was placed beneath and a little to the left of the tract.

Fig. 10.

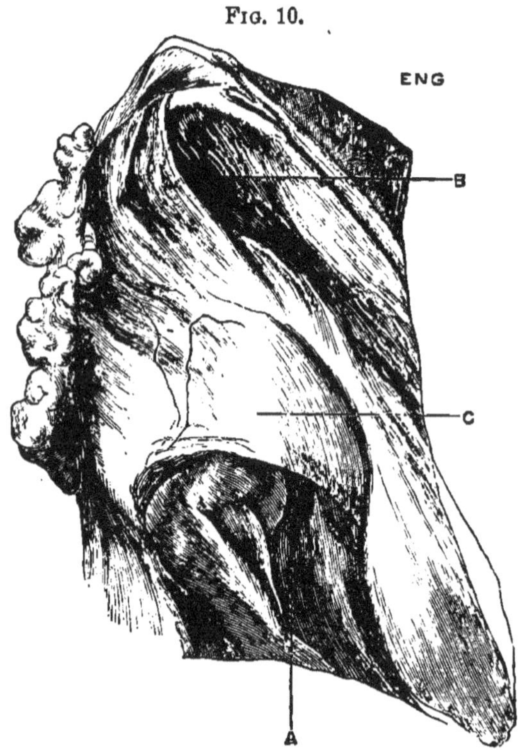

Fig. 10. The septum viewed from Eng's side.
A. The orifice of the umbilical pouch of Eng.
B. The orifice of the hepatic pouch of same.
C. Suspensory ligament of Eng containing the umbilical pouch of Chang.

3. *The vascular structures of the band* were as follows:—

The livers being united, it was found that a colored injection thrown into the portal vein of Chang passed into the liver of Eng. A careful dissection of the bloodvessel (Fig. 11, c) proved it to be a terminal twig of the portal system of Chang. It was of the thickness of a No. seven catheter, French scale, gradually diminished in size, and was lost toward the centre of the band. It did not pass as such across the band, but appeared to break up into minute branches before reaching the liver of Eng. At the same time there was undoubted distension of the portal capillaries with the colored fluid under the capsule of the dorsal surface of the right lobe of Eng's liver, one and one-half inches from the band. Examination of the branches of the mesenteric veins of Eng revealed the curious fact that some of them had received the in-

FIG. 11.

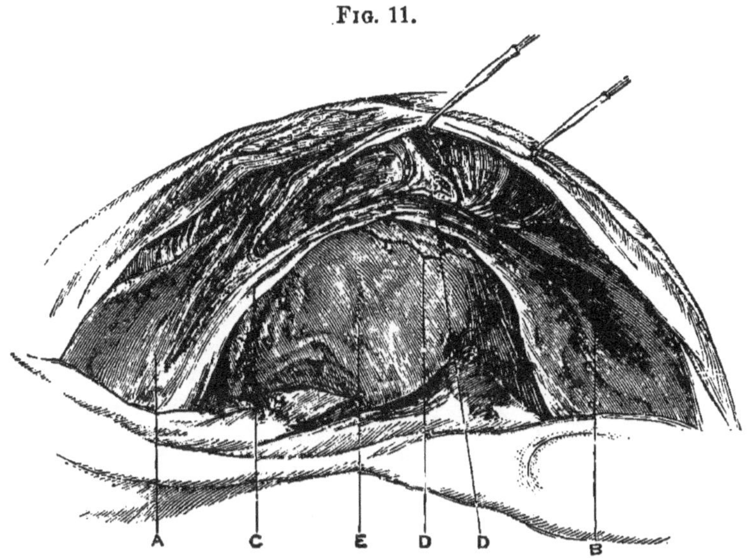

FIG. 11. The surface, C. R., E. L., with pouches removed to display the hepatic tract.

A. Liver of Chang.
B. Liver of Eng.
C. Portal vessel of Chang.
D, D. Minute branches of hepatic artery.
E. Subcutaneous fat of surface, E. R., C. L.

jection. This had not been transmitted through the liver, for the portal vein at the transverse fissure was empty, but through a distinct extra-hepatic portal track, which was found lying under the peritoneum beneath the position of the hepatic pouches, and in association with the umbilical pouches. This vessel began by relatively large radicals towards Chang's side, became larger as these encroached on Eng's side, and was finally received within the portal system of Eng's body, as a tributary to its mesenteric vein.

No other vessels were met with in the band excepting a few insignificant branches of the hepatic artery, and the terminal twigs of the right internal mammary of Eng. The former vessels are marked D, D, Fig. 11. The latter vessel terminated by piercing the diaphragm, and giving ultimate filaments to the integument of the "front" of the band as shown in Fig. 12.

FIG. 12.

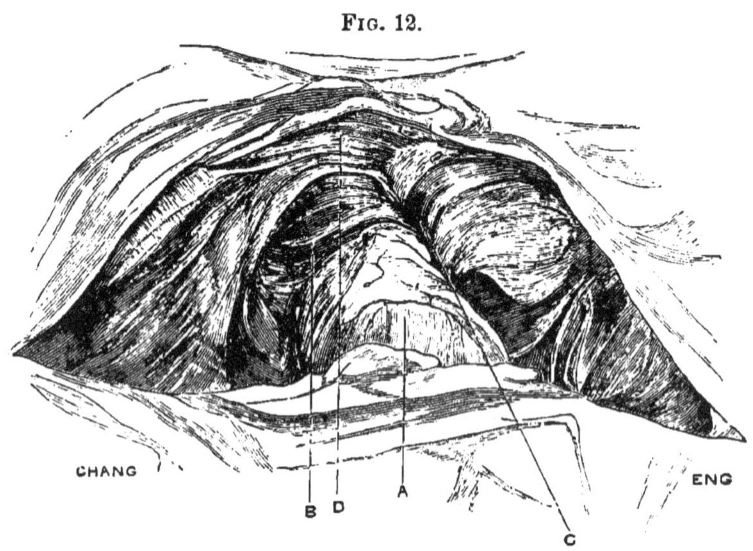

FIG. 12. The surface, C. R., E. L., with pouches, hepatic tract, and peritoneal attachments removed to display the diaphragms.

A. Subcutaneous fat of surface, E. R., C. L.

B, C. Symmetrical muscular fasciculi.

D. Fasciculi of Eng crossing the median line of the band.

4. *The diaphragms.*—The subject being in the same position as in Fig. 2, the livers were removed, the peri-

toneal coverings dissected from the band, and the diaphragms exposed (Fig. 12). The point (A), marked by the terminal twigs of the right internal mammary of Eng, indicated the "anterior" of the band. A broad slip of fibres of Chang (B) was seen to pass across the median line, and to be inserted into the left border of the ensiform cartilage of Eng (Fig. 17). This arrangement would appear to correspond to the smaller collection of fibres (C) belonging entirely to Eng. A second

FIG. 13.

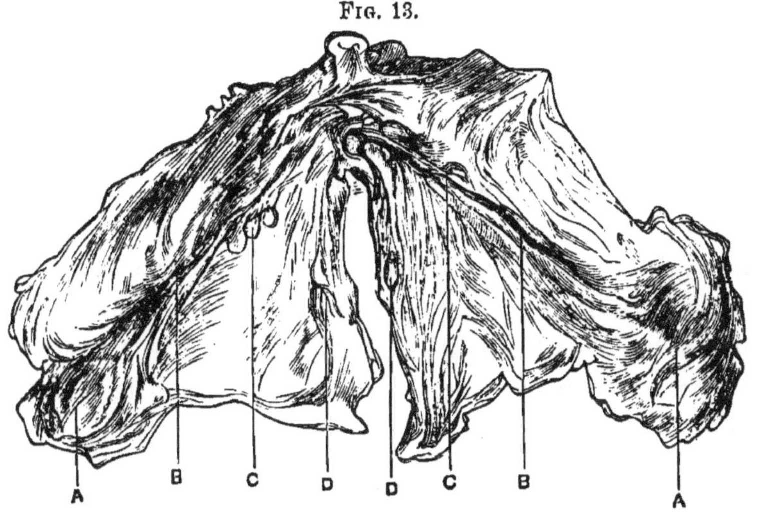

FIG. 13. The peritoneal linings of the anterior walls of both abdominal cavities.

A, A. The summits of the bladders.
B, B. The umbilical ligaments.
C, C. The nodules of fat at the parietal scar.
D, D. The isolated lobules of fat.

arrangement of fibres was seen above those just indicated, immediately under the cartilages (D). This appeared to arise from the border of the cordiform tendon of Eng by two distinct narrow slips, which crossed the median line to be inserted dispersedly on the diaphragm of Chang.

5. *The ensiform cartilages.*—After removing the diaphragms the cartilages were exposed. They may be described as follows:—

CHANG.—The cartilage measured 2½ in. wide, and 8 in. in length along its axis. The right border was very prominent, and projected ½ in. beyond the limit of the corresponding border in Eng. It was almost in close contact with the cartilage of the eighth rib; it was very robust, with upper surface convex, under surface nearly plane. The left lateral border was 2½ in. in length, right lateral border 11 lines in length. The former was marked by three tubercles of about equal size. One situated about 1 in. from the sternal origin; the other about 2 in. from the same point; the third at its extreme anterior border. None of these were robust, or presented any of the thickening noticed on the right side. The middle of these tubercles was on a line with that of the posterior tubercle. The junction of the ensiform process with the sternum was not marked by the eminence characterizing the similar point in Eng.

ENG.—The cartilage differed from that of Chang in being 2⅓ in. wide, 11 lines in length of axis. The left lateral border was abruptly deflected downward, and did not present the transverse smooth projection noticed in Chang. This deflection was almost at right angles to the dorsal surface, acuminate inferiorly, and presenting a straight surface toward Chang, and an oblique one toward the ribs. The length of left lateral border was 1 in. The right lateral border, 1⅓ in. in length, presented a smooth sub-rounded edge without tubercles, and terminated in a free rounded border on a plane a little above that of Chang. On the whole this border was more robust than that of Chang. On the dorsal aspect of the process near its base was seen the rounded eminence described in the account of the external appearances (see page 7).

A comparison between the two ensiform cartilages shows that in Chang the anterior border was longer than in the right in Eng. In other proportions Eng's was equal if not larger than Chang's, and was more robust.

The union between the cartilages was of the character of a symphysis. The union was very intimate along the border E. L., C. R., "posterior" (Fig. 14); the

FIG. 14.

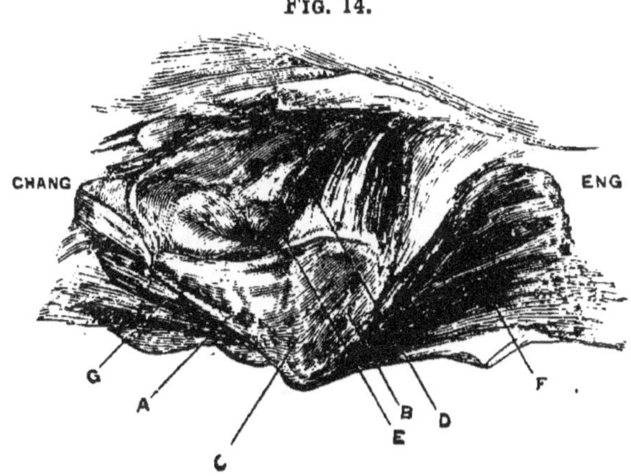

FIG. 14. A section of both ensiform cartilages, C. R., E. L.
- A. Chang's cartilage.
- B. Eng's cartilage.
- C. The synchondrosis.
- D. The bursa-like sac covering the same.
- E. An opening into the sac.
- F. Transversalis muscle of Eng.
- G. Transversalis muscle of Chang.

exposure of the junction by a delicate transverse cut showed a close union between the cartilages, thus constituting this part of the band a synchondrosis. That this, however, did not characterize the entire line of apposition was at once seen by turning to the border E. R., C. L. ("anterior"), where an interval, two lines in width, was seen between the cartilages, an interval which had been evidently susceptible of variation during life. This interval extended across one-fourth the width of the band. The portion of the band between the parts as above indicated, was occupied by a bursa-like sac (Fig. 14, D), which was opened by a minute orifice (E) to display its true nature. This sac was crossed above by a stout band of fibrous tissue (Fig. 15, A) an inch in width. Beneath, the sac was protected by a less well-defined band of the same width as the upper ligament, and which crossed between the two processes, to be lost in the perichondrium.

FIG. 15.

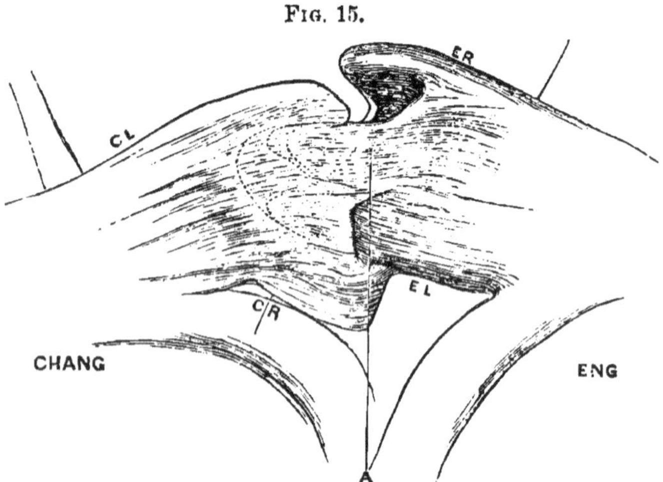

FIG. 15. Upper surface of ensiform cartilages.
A. The upper ligament uniting the cartilages.

Viewing the cartilages as the constituent parts of the band, we found the border C. L., E. R., the "anterior," to be longer than C. R., E. L., the "posterior."

C. L., E. R. was a convex, nearly even border, C. L. being larger than E. R., with a gaping interval placed nearer Eng than Chang. C. R., E. L. was an irregular, uneven border, without interval, C. R. being smaller than E. L., and placed to the outer side.

VI. Other Structures.

1. *The spleens.*—This organ in Eng was 5 in. long, $3\frac{1}{3}$ in. wide. The dorsum was marked by a large sulcus, extending nearly across the organ, continuous with the sulcus on the superior border. The hilus was relatively shorter than that of Chang, beginning above, fully an inch below its upper border, and terminating within a half inch of its inferior border.

In Chang it measured 5 in. long, $2\frac{1}{3}$ in. wide. It was sub-elliptical in form, upper lip somewhat abruptly compressed. The lower border was obtuse and rounded. The dorsum was smooth, and presented at its posterior edge a single sulcus placed midway between the tip and the inferior border. The hilus extended nearly the entire length of the under surface.

2. *The livers* (Fig. 16).—In Eng the liver was 9 in. broad. The right lobe was $7\frac{1}{2}$ in. wide, antero-posteriorly. The fundus of the gall-bladder was seen on the anterior edge of the organ. The only noticeable feature on the under surface of the liver, was the lobus Spigelii. This was large, measuring 2 in. in transverse diameter, and $2\frac{1}{3}$ in. in antero-posterior diameter. It presented a somewhat increased breadth of neck, which was overlapped by an anterior prolongation of the lobe, and terminated by a rounded compressed extremity at the transverse fissure. The quadrilateral lobe was

well developed, 2 in. long in greatest diameter, 10 lines wide.

FIG. 16.

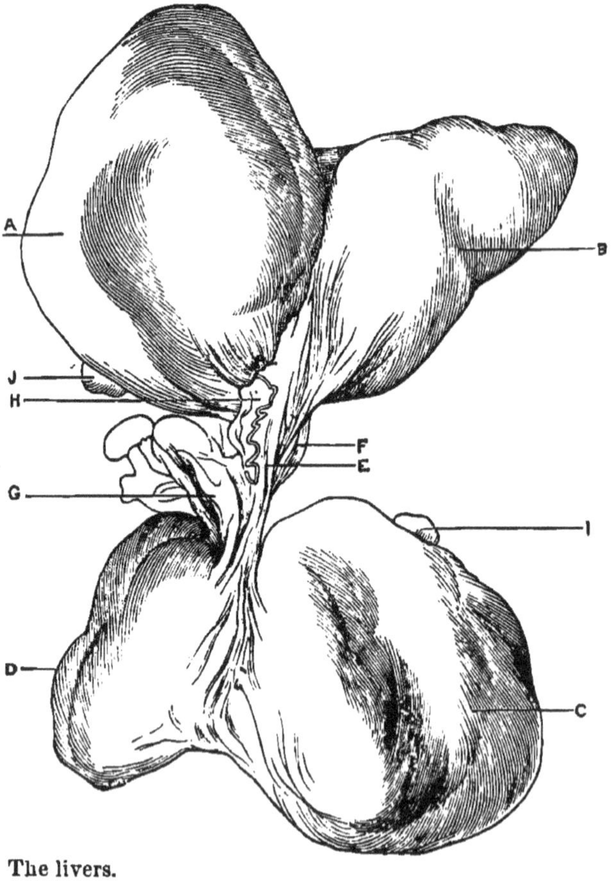

FIG. 16. The livers.

A. Right lobe of Eng.
B. Left lobe of same.
C. Right lobe of Chang.
D. Left lobe of same.
E. Hepatic tract.
F. Round ligament of Eng.
G. Round ligament of Chang.
H. Accessory suspensory ligament of Eng, with termination of the right mammary artery.
I. Fundus of gall-bladder of Chang.
J. Fundus of gall-bladder of Eng.

In Chang the liver was $8\frac{1}{3}$ in. broad. The right lobe was 5 in. wide, antero-posteriorly. The appearance of the gall-bladder corresponded to that seen in Eng. The under surface was normal. The lobus Spigelii presented a narrower neck than in Eng, the

Fig. 17.

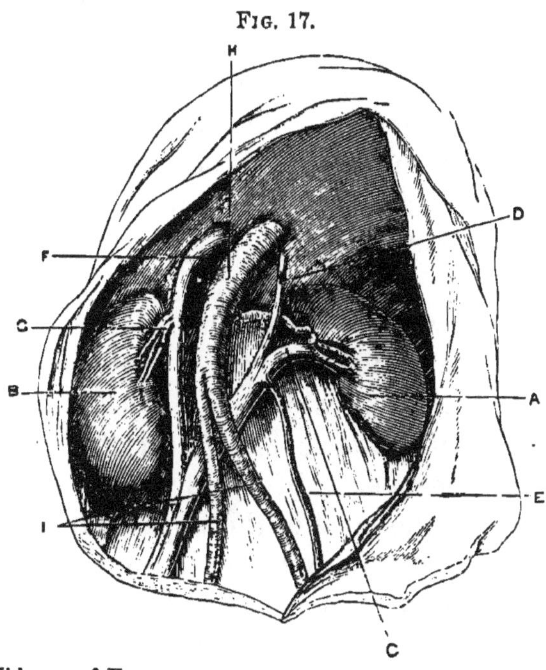

Fig. 17. Kidneys of Eng.
- A. Left kidney.
- B. Right kidney.
- C. Left renal vein.
- D. Left supra-renal vein.
- E. Left spermatic vein.
- F. Descending vena cava not distended with clot.
- G. Right renal vein.
- H. Aorta distended with plaster.
- I. Primitive iliac arteries.

anterior prolongation being greater. The quadrilateral lobe was less developed than in Eng. Indeed it was not raised above the under surface of the right lobe, and its limits were so imperfectly marked that it could not well be measured.

3. *The kidneys.*—In Eng, the body lying on the table E. R., C. L., the left kidney (Fig. 17, A) was 4 in. long, $1\frac{1}{2}$ wide at its hilus, and of the usual kidney shape. It lacked $\frac{1}{2}$ in. of reaching the crest of the ilium. The renal vein (Fig. 17, C) of the same side measured 3 in. in length, and was decidedly oblique in position, its termination in the cava being below the level of the lower end of the kidney.

The right kidney (Fig. 17, B) corresponded in position to the left kidney of Chang, that is to say, it was in the shallower portion of the abdomen, and in contact with the abdominal wall. It measured 4 in. in length, and $2\frac{1}{4}$ in. in width. Its inferior border lacked 2 in. of reaching the superior crest of the ilium. The renal vein ascended a little upward to enter the cava a little below the level of the upper end of the kidney.

In Chang, the body lying in such a way that the great trochanter of the right side rested on the table, the left trochanter being raised three inches from the same plane, an obliquity was given to the trunk, and rendered the position of the abdominal organs somewhat anomalous.

The left kidney (Fig. 18, A) lay with its lower half clearly within the iliac fossa, its inferior border answering to a point an inch and a half below the termination of the aorta. The organ lay, at its inner and inferior portion, upon the left primitive iliac vein; it measured $3\frac{3}{4}$ in. in length, and $2\frac{7}{12}$ in. in width at its widest part. It was larger below, where it retained the usual appearance, but was somewhat abruptedly pointed above, and was marked by the characteristic notch on its inner side. The renal vein (Fig. 18, C) was very obliquely situated, indeed was almost parallel with the cava, and was $3\frac{1}{2}$ in. long. The termination of the renal vein answered to a line running across the abdomen lying fully 1 in. above the upper end of the left kidney.

FIG. 18.

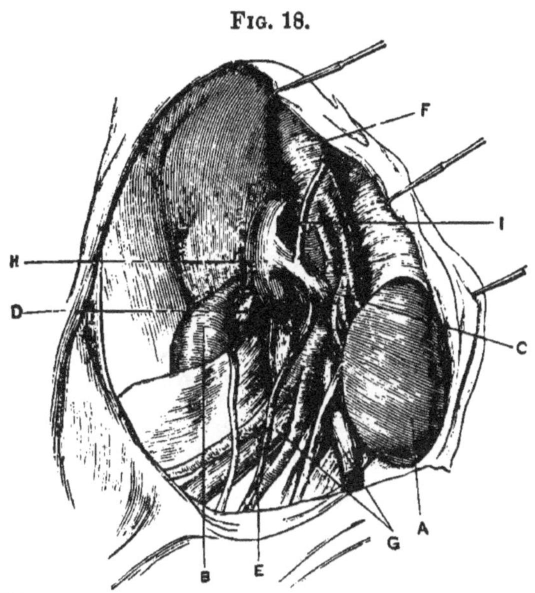

FIG. 18. Kidneys of Chang.
- A. Left kidney.
- B. Right kidney.
- C. Left renal vein.
- D. Right renal vein.
- E. Left spermatic vein.
- F. Aorta filled with plaster.
- G. Primitive iliac veins.
- H. Descending cava distended with clot.
- I. Left supra-renal vein.

The right kidney (Fig. 18, B) was normally situated. It measured 4 in. in length, and $1\frac{1}{2}$ in. in width at its centre, and presented the usual reniform appearance.

Its inferior edge just reached an eminence answering to the superior crest of the ilium.

4. *The testicles.*—The right testicle of Eng was normal. The left testicle was not within the scrotum. Dissection from within the abdomen showed that the organ had been retracted. It lay well concealed within the inguinal canal, slight traction making it appear within the abdomen.

The testicles of Chang were normal.

5. *The hearts.*—The heart in Eng was situated nearer the median line than normal. The abdominal incision was very unfavorable for studying its exact position in the mediastinum. It was removed through an opening made in the diaphragm. The right side of the heart was occupied by a soft grumous clot much smaller than was found in the same locality in Chang, and which did not distend the cavities. The left side was normal. It was without clot so far as could be determined in the injected condition of the ventricle.

The heart of Chang presented a right auricle and ventricle distended with a dense venous clot; this extended from the right ventricle along the pulmonary arteries. The left side of the heart was empty.

The ductus arteriosus and foramen ovale were firmly closed in both hearts.

6. *The vessels.*—The arteries of both subjects were, so far as examined, in an extremely atheromatous condition. Large plates of calcareous matter were deposited in the abdominal aortas. The injecting matter flowed insufficiently in the left lower extremity of Chang, from a clot plugging the femoral artery.

The venous system of Chang was engorged, giving the appearance of these vessels having been injected after death; that of Eng was comparatively empty.

7. *The lungs.*—The lungs were so altered by *post-mortem* changes prior to embalming, their contraction by the chloride of zinc, and their increase of weight from the plaster, that no extended examination was made of them. But little difference was seen between the conditions of the lungs in the two men. No hepatization was present in Chang.

8. *The vertebral column and ribs.*—There was marked lateral curvature of the vertebral column in both bodies. This was more conspicuous in Chang. The convexity of the curve was about half-way down the vertebral column, and inclined in Chang to the right side. The distance from the centre of the vertebral column to the left abdominal wall, 2 in.; to the right abdominal wall, 5 in. The left side of the abdominal cavity, measuring from about the level of the band to the last rib of the right side, 7½ inches.

The ribs in both Chang and Eng were 22 in number, 7 true and 4 false. On the right side of Eng the first, second, and third ribs were normal. The fourth, fifth, sixth, and seventh presented diminished intercostal spaces, owing probably to the extreme traction made on them by the deflection of the ensiform cartilages. The intercostal space between the third and fourth ribs was slightly contracted; that between the fourth and fifth ribs was very much contracted, the muscle being bulged inward. Between the fifth and sixth, and sixth and seventh ribs the space was less contracted. The remaining intercostal spaces were about normal. The fifth rib near its articulation with the vertebral column formed a well-defined ridge within the thorax, carrying with it the sixth and seventh ribs, thus forming a rounded elevation, distinguishing the positions of these ribs from the thoracic wall above and below this point, where the parietal surface presented the usual concave appearance.

On the left side of Chang a similar arrangement of ribs and intercostal spaces was seen to the above.

The remaining organs were not examined.

REMARKS.

With reference to the cause of death of the Siamese twins it may be briefly said that, in consequence of the restrictions by which we were bound, no examination of the brains was made. It cannot, therefore, be proved that the cause of Chang's death was a cerebral clot, although such an opinion, from the suddenness of death, preceded as it was by hemiplegia and an immediate engorgement of the left lung, is tenable. Eng died, in all probability, in a state of syncope induced by fright—a view which the over-distended bladder and the retraction of the right testicle would appear to corroborate.

The existence of lateral curvature was not unsuspected. It was known to those who had examined the twins before death. Indeed, it must have been a necessity of the acquired position.

The presence of a pad of subperitoneal fat at the usual position of the umbilicus was certainly curious.

It would appear to be an example of a localized nutritive change about the peritoneum, at the centre of the umbilical region, anticipating the exit of the vessels of the cord at that point. Familiar examples of this anticipation between structures developing from different layers of the embryo are seen in malformations of the genital organs, eye, ear, etc. In the above example it is remarkable only from the rarity of the conditions yielding it.

The circulation in each individual of the twins was practically complete, since the demonstration of continuity between the portal systems, although satisfactory, invites the conclusion that the amount of blood which passed from one to the other side of the band must have been, in the condition of the parts at the time of the demonstration, very inconsiderable, and was not competent in all probability to modify the performance of any act of the economy.

In the fœtal and early period of extra-uterine life the vessels must have been more capacious, and associated with a large tract of liver tissue. It follows, all things being equal, that an attempt at division of the band in early life would have been accompanied with more venous hemorrhage than at any subsequent period.

In proof that the twins were the product of a single conception, the strict correspondence between the markings of the two spleens, as well as the number of the ribs, may be observed. The absence of available data bearing upon the question of symmetry between visceral organs of twins, prevents us from drawing here too positive an inference. It is probable, however, that the twins were individuals of a single organism, remarkable for its complete expression of duplex bilaterality.[1]

FIG. 19.

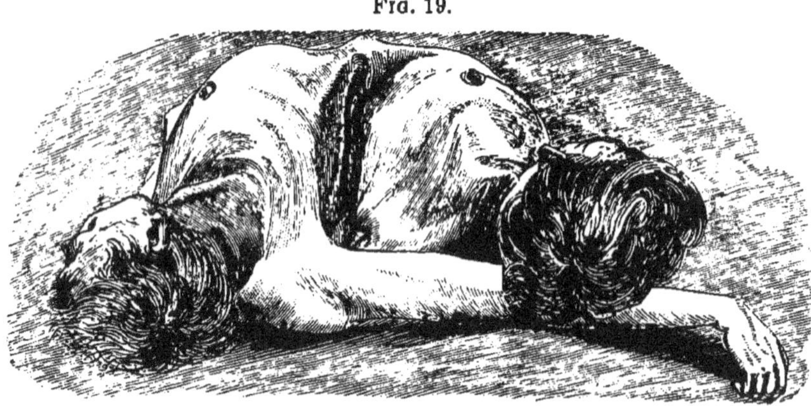

FIG. 19. Foreshortened view of the trunks, showing in the acquired position the band from above and the contours of its lateral surfaces.

---

[1] I desire to return thanks to Dr. T. H. Andrews and Dr. J. W. White, Jr., for important assistance rendered in preparing the notes of the autopsy.

DESCRIPTION OF FIGURES.

FROM PHOTOGRAPHS.

FIG. 1. Twins in acquired position (E. R., C. L.). Taken in St. Petersburg, 1870. Page 3.

FIG. 2. Twins in acquired position (E. R., C. L.). Taken after death at Philadelphia. Page 8.

FROM SKETCHES.

FIG. 3. The surface, C. R., E. L., exposed by removal of skin and superficial fascia to display the tendons of the external oblique muscles and adjacent parts. Page 12.

A. The superficial fascia—lost over the position of Chang's umbilical pouch.

B, C. Supplemental layers of fibrous tissue of Eng not seen in Chang; B is a continuation toward Eng of aponeurotic fibres having a source from the linea alba of Chang; C is independent of the former, and is continuous with the deep pectoral fascia.

D. The interlacing of fibres on tendon of external oblique muscle of Chang.

E. The linea alba of Chang, beginning at C. R.

F. Its continuation to E. L., and insertion upon the ensiform cartilage.

FIG. 4. The umbilical ligament in Eng. Page 15.

A. The umbilical ligament.

B. The lobule of fat at position of normal umbilicus.

FIG. 5. The umbilical ligament in Chang. Page 16.

The letters as in Fig. 4.

FIG. 6. The abdominal organs of Eng—the small intestines removed. Page 17.

A. Left lobe of liver.

B. Right lobe of liver.

C. Gall-bladder.

D. Suspensory ligament.
E. Lobules of fat in the position of the termination of the umbilical ligament.

FIG. 7. The abdominal organs in Chang—the small intestines removed. Page 19.
The letters as in Fig. 6.

FIG. 8. The surface, C. R., E. L., showing the interior of band by free division of the aponeuroses seen in Fig. 7, and their underlying peritoneal attachments. Page 24.
A. The orifice of umbilical pouch of Eng.
B. The orifice of umbilical pouch of Chang, showing connection with suspensory ligament of Eng.
C. The fenestrated umbilical pouch of Eng passing between the folds of the suspensory ligament of Chang.
D. Suspensory ligament of liver of Eng.
E. Hepatic tract.
F. Hepatic pouch of Eng
G. The septum.

FIG. 9. The septum viewed from Chang's side. Page 25.
A. The orifice of umbilical pouch of Chang.
B. The orifice of hepatic pouch of Chang.
C. Suspensory ligament of Chang, containing umbilical pouch of Eng.

FIG. 10. The septum viewed from Eng's side. Page 26.
A. The orifice of umbilical pouch of Eng.
B. The orifice of hepatic pouch of Eng.
C. Suspensory ligament of Eng, containing umbilical pouch of Chang.

FIG. 11. The surface, C. R., E. L., with pouches removed to display the hepatic tract. Page 27.
A. Liver of Chang.
B. Liver of Eng.
C. Portal vessel of Chang.
D, D. Minute branches of hepatic artery.
E. Subcutaneous fat of surface, E. R., C. L

FIG. 12. The surface, C. R., E. L., with pouches, hepatic tract, and peritoneal attachments removed to display the diaphragms. Page 28.

A. Subcutaneous fat of surface, E. R., C. L.
B, C. Symmetrical muscular fasciculi.
D. Fasciculi of Eng, crossing median line of band.

FIG. 13. The peritoneal linings of the anterior walls of both abdominal cavities. Page 29.

A, A. The summits of the bladders.
B, B. The umbilical ligaments.
C, C. The nodules of fat at the parietal scar.
D, D. The isolated lobules of fat.

FIG. 14. A section of both ensiform cartilages, C. R., E. L. Page 31.

A. Chang's cartilage.
B. Eng's cartilage.
C. The synchondrosis.
D. The bursa-like sac covering the same.
E. An opening into the sac.
F. Transversalis muscle of Eng.
G. Transversalis muscle of Chang.

FIG. 15. Upper surface of ensiform cartilages. Page 32.

A. The upper ligament uniting the cartilages.

FIG. 16. The livers. Page 34.

A. Right lobe of Eng.
B. Left lobe of same.
C. Right lobe of Chang.
D. Left lobe of same.
E. Hepatic tract.
F. Round ligament of Eng.
G. Round ligament of Chang.
H. Accessory suspensory ligament of Eng, with termination of the right mammary artery.
I. Fundus of gall-bladder of Chang.
J. Fundus of gall-bladder of Eng.

FIG. 17. Kidneys of Eng. Page 35.

A. Left kidney.

B. Right kidney.
C. Left renal vein.
D. Left supra-renal vein.
E. Left spermatic vein.
F. Descending vena cava not distended with clot.
G. Right renal vein.
H. Aorta distended with plaster.
I. Primitive iliac arteries.

FIG. 18. Kidneys of Chang. Page 37.
A. Left kidney.
B. Right kidney.
C. Left renal vein.
D. Right renal vein.
E. Left spermatic vein.
F. Aorta filled with plaster.
G. Primitive iliac veins.
H. Descending cava distended with clot.
I. Left supra-renal vein.

FIG. 19. Foreshortened view of the trunks, showing in the acquired position the band from above, and the contours of its lateral surfaces. Page 42.

www.ingramcontent.com/pod-product-compliance
Lightning Source LLC
Chambersburg PA
CBHW022024190326
41519CB00010B/1592

* 9 7 8 3 3 3 7 2 4 8 9 6 3 *